D1548148

To my family and friends

CONTENTS

FOREWORD

BY HOPE TAFT

Seldom does one have the pleasure of being asked to contribute a foreword to a wonderful collection of essays on botanical topics.

Ten years ago I would not have been considered for this honor. I knew very little about Ohio's botanical treasures or where to find them. But I was blessed to have some very knowledgeable guides to literally and figuratively take me around the state and educate me on the many natural wonders it has to offer. I was so inspired by the beauty of its native plants and the stories about their uses and their relationship to geology, weather patterns, industry, and animal life that they can tell, I wanted to share my new knowledge with others. Thus, the Heritage Garden at the Ohio Governor's Residence was started in 2000 and today flourishes as a place to take a walk "on the back roads of Ohio" and discover these special plants in the essence of their natural physiographic regions. Ohio is blessed with five of these regions and thus has a wide variety of plants living on the edges of their preferred habitats. This, in turn, has provided an ample variety of wildlife seeking specific habitats to fulfill special needs in the continuing coevolution of plants and animals.

I quickly learned who the real experts were on the botanical character of various parts of the state. Since each of its four corners and the central area has its own different plant communities, it was difficult to find those knowledgeable on all the vascular plants of the state. One such expert is the author of this book as well as the "Bible" of Ohio plants. Tom Cooperrider came into my life when I was given a copy of his book *Seventh Catalog of the Vascular Plants of Ohio* and I sent him a request to quote from it for an article on the Heritage Garden. I have been an avid student of his ever since. You, too, will learn a great deal about Ohio and its plants by reading this book of essays he wrote over a period of years.

In addition to being an expert on the flora around Kent, Ohio, Tom is also a towering figure among the state's botanists. A number were once his graduate students. His inventories of Ohio plants helped create the basis of the state's rare plant list. He is an expert on Ohio's flora and was instrumental in getting the Ohio Natural Areas Act of 1970 passed, which led to the creation of a statewide preserve system by the Ohio Department of Natural Resources. One of the department's first purchases with its income tax check-off funds was the land now known as the Tom S. Cooperrider–Kent Bog State Nature Preserve.

By turning the pages of this book and absorbing Tom's knowledge from its beautifully written passages, you will not only discover local history and fascinating information about plants but also be drawn into the world of a botanist and have its secrets revealed in language every gardener can understand. Tom Cooperrider continues to teach and influence the course of Ohio's floristic history.

HOPE TAFT
First Lady of Ohio, 1999–2007

PREFACE AND ACKNOWLEDGMENTS

These collected essays deal with my work on the Ohio flora and my botanical activities and observations in and near the city of Kent, Ohio. They center on the years from 1958 to 1993, when I was a faculty member at Kent State University. They also cover related work of colleagues and associates during those years.

The introductions to each chapter are intended to make the essays more accessible to nonbotanists and to place the local activities and observations in larger state and national contexts. Each essay is written as a stand-alone piece.

An earlier version of the first two essays, "On the Flora of Ohio" and "The Study and Conservation of the Ohio Flora," appeared in the *Denison Journal of Biological Science* (Fall 1999). Earlier versions of five other essays were published in *On the Fringe: Journal of the Native Plant Society of Northeastern Ohio:* "The Herrick Magnolia Gardens" (September 2007); "American Holly, George Washington, and 'Beautiful Ohio'" (March 2002); "Kent Bog State Nature Preserve" (September 2004); "The Seasons at the Kent Bog State Nature Preserve" (December 2006); and "The Work of a Twentieth-Century Kent Botanist" (September 2008). Also, versions of the five sections of "The Seasons at the Kent Bog State Nature Preserve" were published separately in 2005 and 2006 in Portage County's daily newspaper, the *Record-Courier.*

The photographs, like snapshots in an album, illustrate some of the plants, places, and people mentioned in the essays. The photograph of tamarack trees on p. 70, appeared in the *Record-Courier,* November 5, 2005.

I am honored to have the foreword for this book written by Hope Taft. The magnificent Heritage Garden she created at the Ohio Governor's Residence is a tribute to the complexity and beauty of the Ohio flora.

I am also honored to have the afterword written by David E. Boufford, senior scientist at the Harvard University Herbaria. His studies on the flora of eastern Asia are a model for floristic workers everywhere, and his fieldwork in China and Japan the envy of many, myself included.

My thanks go to the photographers, whose work enhances the book: Dick Bentley, Ray Black, Dave Brandenburg, Bob Christy, Mix Cooperrider, Sue Cooperrider, Guy Denny, Gary Harwood, Margaret Popovich, Al Schleider, and Gerry Simon. My thanks go also to the staff at Campus Camera in Kent.

I am grateful to members of my family for their help. My wife, Miwako, gave the manuscript a critical review from the perspective of a botanist and participant in many of the events described. My sister, Sue, reviewed the manuscript from a writer's (and nonbotanist's) perspective and assisted in research matters. My daughter, Julie, helped with the indexing, and my son, John, made the special container for the Davey specimens and the drawing of large white trillium.

I thank all the graduate students for their work in the Kent State program and for their contribution to increased knowledge of the plants of Ohio. The following students earned graduate degrees in the program described in Chapter 3, each completing a research project that involved, entirely or in part, some aspect of the Ohio flora: Joyce E. Amann, Sandra Mutz Anderson, Barbara K. Andreas, Gregory D. Bentz, Judy S. Bradt-Barnhart, Bruce L. Brockett, James F. Burns, Robert J. Cline, Tammy E. Cook, Virginia L. Curtis, Allison W. Cusick, David P. Emmitt, Edward J. P. Hauser, William Dean Hawver, George A. McCready, Lynne D. Miller, Paul L. (Larry) Pusey, Robert F. Sabo, Gene M. Silberhorn, Lewis W. Tandy, Paula L. Van Natta, and Hugh D. Wilson. Judith Doty Dumke completed an important nondegree post-master's study.

I also thank the many undergraduate students who worked on some aspect of the floristics of Ohio or on that of more distant places or on herbarium or laboratory tasks during their years at Kent State. Their contributions were of inestimable value.

There are many people to thank for their role in making this book possible: Will Underwood, director of the Kent State University Press, and the talented KSU Press staff, for preparing this book for publication; Carolyn Platt, reader for the KSU Press, for her valuable comments and suggestions; Roger Di Paolo, who reviewed the manuscript with special attention to Kent history; Gerald Shanley, Kent city arborist, and Heidi Hetzel-Evans, Charlotte McCurdy, and Emliss Ricks Jr., staff members of the ODNR's Division of Natural Areas and Preserves, all of whom provided detailed information and help with specific problems; Gina Flick, of the KSU Foundation and Development staff, who made arrangements for the photograph of Japanese magnolias (p. 40); Ann Malmquist and Jane McCullam, editors of *On The Fringe;* Michael Barker and Sarah Emery, student members of the 1999 editorial staff of the *Denison Journal of Biological Science,* and their faculty adviser Julianna Mulroy; and Guy Denny, retired chief of the Division of Natural Areas and Preserves, for his support and help over many years.

Most of all, I thank Linda Matz, who typed and processed the manuscript through many revisions. I did not even begin work on the project until I was certain of her availability and willingness to take on the task. I appreciate her skill and unfailing professional courtesy.

INTRODUCTION

The second half of the twentieth century marked a turning point in our thinking about the earth. Around the globe, people took to the streets to protest a number of environmental problems, their actions associated with the rise of the modern environmental movement.

In the United States, the environmental movement dates from the establishment of Yellowstone National Park in 1872 and the founding of the Sierra Club in 1892. It emerged as a cohesive movement in the 1960s, propelled forward by two notable events. In 1962, American biologist Rachel Carson published *Silent Spring*, a book warning of the harm being done to "the fabric of life."[1] And, in 1968, American astronaut William Anders, while orbiting the moon, shot *Earthrise*, a photograph of the earth ascending above the lunar horizon. There, for all to see, was our home, a blue planet with a swirl of white clouds. Anders later spoke of "the beauty of Earth—and its fragility."[2] The words of Carson and Anders were soon applied to many specific parts of the environment, including the earth's plant life.

Most of the concern about plant life was in the area of floristic botany (floristics), the branch of botanical science dealing with the identity and frequency of the plant species of a particular area—in other words, what species grow in the area and how common or rare each species is. There was a call from floristic botanists for the preservation of wild natural areas and

Tamarack tree at Kent Bog State Nature Preserve (photograph by Gerry Simon)

the protection of rare native plant species. New federal and state legislation enacted during the 1960s and 1970s brought these botanical goals into the mainstream of contemporary life.

The federal Wilderness Act of 1964 established "a national wilderness preservation system for the permanent good of the whole people," to secure

Small cranberry, one of Ohio's rarest plants, now protected at Kent Bog State Nature Preserve (Photograph by Dick Bentley)

"the benefits of an enduring resource of wilderness." Following suit, the Ohio Natural Areas Act of 1970 established a statewide system of nature preserves to be administered by the Ohio Department of Natural Resources (ODNR). Stemming from the Ohio act was the purchase, in 1985, of the local land that was to become the Kent Bog State Nature Preserve.

In the federal Endangered Species Act of 1973, rare species were declared worthy of protection because of their "esthetic, ecological, educational, historical, recreational, and scientific value to the Nation and its people." The Ohio Endangered Plant Law of 1978 assigned to the ODNR's Division of Natural Areas and Preserves the task of establishing and maintaining an up-to-date list of rare native plant species and the responsibility for making rules and regulations providing for their protection. Tamarack (*Larix laricina*), tawny cotton-grass (*Eriophorum virginicum*), and small cranberry (*Vaccinium oxycoccos*), all of which grow at the Kent Bog State Nature Preserve within the city limits of Kent, are some of the rare species included in that list.

It was a time of focusing increased attention on all plants, cultivated and wild. As attitudes and legislation affecting the plant life of the United States and the state of Ohio were undergoing revision and change, interest in the care and cultivation of plants in the city of Kent was also on the rise.

The interest, however, was longstanding. The most admired gardens in the city's history were those of Daisy Wolcott at the family home on West Main Street. During the 1930s, when the Wolcott gardens were in their prime, they included seventy-five varieties of lilacs and seventy varieties of peonies and attracted thousands of visitors each spring.[3]

The Kent Environmental Council, formed in 1970, led a successful campaign to clean up the Cuyahoga River and its banks and to form new city parks bordering the river, with paths among the wild plants growing there as well as among the newly planted trees and shrubs. In neighborhoods throughout the city, more and more homeowners maintained front-yard flower beds and backyard vegetable gardens. The Beckwith family's peach and apple orchard east of town became a favorite site for outings and the purchase of fresh fruit.

In 1985, the National Arbor Day Foundation granted Kent the title of "Tree City USA" for its demonstrated commitment to the planting and caring for trees in the community. Carol Lockhart of Kent's public service department and Fred Simons of the shade tree commission led the application effort for this honor. Kent's inclusion in this group of American cities was especially fitting because it is home to the headquarters of the Davey Tree Expert Company, whose world-famous school for training tree surgeons was established in Kent in the early 1900s.

Underlying the federal, state, and local activity brought on by the modern environmental movement was a growing understanding that natural areas are important elements in stabilizing the environment, that rare and endangered plants are important components of strong ecosystems, and that access to plants—both wild and cultivated—is important for the human environment and the human spirit.

CHAPTER 1

THE LOCAL FLORA

Members of every human society have studied the plants of their environment, the local flora. On the one hand, these plants often provided the necessities of life, such as food, medicines, materials for buildings and clothing, and fuel for fire. On the other hand, some local plants were poisonous or otherwise harmful and required prompt identification. Given the role of plants in these basic aspects of life and survival, the human drive to study the local flora may be innate. There may well be a "local flora gene" in our makeup.

In the United States, the concept of the *local* flora is often extended to mean that of the individual states. For example, the local Ohio flora encompasses all the wild plants of the state, including the trees and shrubs of the forests; the wetland plants of bogs, fens, and marshes; the grasses of the prairies; the wildflowers and ferns of woodlands and fields; and the weeds of wastelands and other disturbed areas—all of them growing outside cultivation and without human assistance.

These plants are all vascular plants, the group that includes most of the familiar plants of the landscape. Vascular plants are distinguished by the presence of conducting (vascular) tissues that transport the fluids necessary for life throughout the plant body. Nearly all the plants mentioned in this book are vascular; only the algae, fungi, lichens, and mosses are nonvascular.

Botanists working in the state for the past 200 years have collected specimens of some 2,700 different species of vascular plants. Of that total, about 1,800 species are native to Ohio, and about 900 species are alien.[1]

The following essay begins with a recollection of the primeval forest of Etna Township. It is relevant to the botany of the Kent area because Etna Township in Licking County and Franklin and Brimfield townships in Portage County, where Kent is located, all belong to the same Ohio eco-region. The primeval forest of the Kent area was much like that of Etna Township.

On the Flora of Ohio

Until the late 1700s, a vast old-growth forest composed of about 100 species of deciduous trees and a few species of needle-bearing evergreens covered nearly all of Ohio. The trees, many 500–600 years old, were tall and full, their high, dense canopy of summer leaves shading and darkening the ground beneath them.

A Noble Primeval Forest

Morris Schaff left a firsthand description of the original forest of Etna Township in Licking County, Ohio, when, in later life, he recalled his childhood there in the 1840s. Like other observers, he remembered trees so tall that their lowest limbs were fifty or more feet above the ground.

> When I was a boy three fourths of Etna Township was covered by a noble primeval forest. And now, as I recall the stately grandeur of the red and white oaks, many of them six feet or more in diameter, towering up royally fifty and sixty feet without a limb; the shellbark hickories and the glowing maples, both with tops far aloft; the mild and moss-covered ash trees, some of them over four feet through; the elms and sturdy beeches, the great black walnuts and the ghostly-robed sycamores, huge in limb and body, along the creek bottoms, I consider it fortunate that I was reared among them and walked beneath them. . . .[2]

Bur oak, about 500 years old, Goll Woods State Nature Preserve, Fulton County, near Toledo. A remnant of Ohio's primeval forest. (Photograph by Guy Denny)

> I wish there was some way that a just idea could be transferred to this page of the splendor of those woods, when on every hand there rose those stately oaks, ashes, sycamores, and black walnuts, all lifting their heads like kings far up into the sky.[3]

The forest was also home to a number of shade-tolerant shrubs, vines, and ferns living beneath the trees. And during the early part of the growing season, in April and May, a host of wildflowers, probably about 150 species, appeared on the forest floor, most completing their flowering cycle before the trees' broad leaves expanded above them, cutting down the light.

Approximately 95 percent of Ohio was forested. The other 5 percent was more open and included prairie patches, glacial lakes, tamarack bogs, alkaline fens, marshes, fields of sand, rocky cliffs and ledges, and rivers and their muddy banks, each habitat with a distinct flora.[4]

THE NATIVE SPECIES

In all the habitats combined, both wooded and open, the native Ohio flora totaled about 1,800 species of vascular plants. They varied in abundance from common species, such as red maple and large white trillium, to rare ones, such as tamarack and small cranberry.

The wooded land in Ohio was part of a more extensive forest that covered most of the eastern United States.[5] One can imagine that the Native Americans, whose ancestors had lived in the forest for centuries, would have thought the trees a constant, an eternal element of nature. But the European settlers of the early nineteenth century cleared the land for farming in their push through the wilderness. The magnificent, tall trees were cut, and most were simply burned at or near the site where they fell.[6]

Today a few remnants of the original forest survive, each giving us a glimpse of primeval Ohio. One of the finest examples is the old-growth forest at Goll Woods State Nature Preserve in Fulton County, near Toledo. Its bur oak trees are 500 years old, and many measure over four feet in diameter. Isolated individual trees from the primeval forest also survive at scattered locations. For example, the national champion cucumber magnolia tree, estimated to be 430 years old, grows in a North Canton residential area twenty miles south of Kent, in Stark County. Starting from a seed that germinated in about 1580, this grand old tree has a diameter greater than seven feet, a dimension often encountered in the original forest.[7]

Remnant areas of the nonforested habitats, such as marshes, bogs, and prairies, also survive in nature preserves around the state.[8] However, these few surviving areas are not enough to save all the original plant species. Probably some fifty to seventy-five species of vascular plants are permanently lost from the state's flora, most because of habitat destruction.

ALIEN SPECIES ENTER THE FLORA

Scores of alien plant species, mostly Eurasian in origin, moved rapidly into the disturbed habitats left by the clearing of the forests in the nineteenth century. Since then, alien plants have continued to enter and become part of the local flora. At last count, specimens of some 900 species of nonnative vascular plants, growing outside cultivation, have been collected by botanists working in Ohio.[9] About 500 of the alien species are naturalized, that is, firmly established and reproducing and maintaining themselves in the flora year after year. The other 400 alien species have thus far proved to be short lived and have not become established members of the flora.

Many of the naturalized aliens are benign, occupying disturbed sites where native species cannot survive and therefore doing little direct harm to the native flora. An example is the Chinese white mulberry tree, also called silkworm mulberry, that was widely planted in the Kent area in the late 1830s.[10] The mulberry's leaves were intended to provide food for silkworms, part of an attempt to establish a silk industry in Kent. The silkworms did not survive, however, and the attempt failed, but the mulberry trees persisted and became naturalized. White mulberry trees, likely descended from those planted in the 1830s, can be found today in disturbed weedy areas along the Cuyahoga River in Kent.

Tree-of-heaven, also a Chinese tree, is another relatively benign alien. It was introduced into North America in 1784.[11] It quickly spread from cultivation and became widely naturalized across the continent. Tree-of-

Leaves of white mulberry, also called silkworm mulberry, at Tannery Park near Cuyahoga River in Kent (Photograph by Dick Bentley)

heaven now grows in every county in Ohio and at several places in the Kent area, inhabiting disturbed ground where native plants do not prosper.[12]

Other aliens, able to invade natural areas and compete with native plants, are a different matter. A European plant, garlic mustard, is an example of such an invasive species. It is a frequent weed along rural roadsides in the Kent area, where it does no damage to the native flora. But it also invades floodplain forests, where it is extremely harmful, displacing the native species of these habitats. Local conservation groups sponsor garlic mustard "pulls" along the Cuyahoga River in Kent to keep the population size of this invasive alien under control.

FLORAS

We live in the midst of three floras. The plants of the *cultivated flora,* those we intentionally add to our environment, range from lawn grasses and dooryard ornamentals to the plants of gardens and agricultural croplands. The *local flora* includes all the plants that grow wild, outside cultivation, a mixture of native and alien species in places such as woodlands, wetlands, and weed lots. The *local flora of the past* is a special part of the local flora seen in the stands of mature, native plants in nature preserves. These surviving areas from the past show us the floristic landscapes our ancestors saw.

CHAPTER 2

THE FLORISTIC BOTANISTS

Floristic botanists are students of the earth's flora, and they are also its watchdogs. They deal with the world's plants, from tropical rainforests to the arctic tundra, and from remote oceanic islands to urban nature preserves, observing the plants in their natural habitats, collecting specimens to verify what they have seen, and writing articles and books that list and describe the plants of a particular area. When they find species whose numbers are declining and are threatened with extinction, they sound the alarm to let the world know of the danger. For floristic botanists, all plant species are valuable, and each is an important part of the earth's ecology.

The Study and Conservation of the Ohio Flora

In the late eighteenth and early nineteenth centuries, at about the same time European settlers were arriving in Ohio in great numbers, professional botanists began collecting plant specimens in Ohio for scientific study.[1] For most species, these early historical specimens confirm their longtime presence in the state flora. But for a few species, now lost from the flora, they provide the only tangible evidence we have of their past occurrence in the state.

Labrador-tea in a Portage County bog, a northern species, extremely rare in Ohio. (Photograph by Guy Denny)

Sometimes in current fieldwork a species not seen in Ohio for many years is found again in a new location. One such story is that of a northern bog plant whose scientific and common names are linked with both Greenland and Labrador. Labrador-tea (*Ledum groenlandicum*), a member of the heath family, is a small boreal shrub that migrated south into Ohio ahead of the last glaciation of the Ice Age. When I came to the Kent area in 1958, Labrador-tea was last known in Ohio by a specimen collected from Ashtabula County in 1928. On July 11, 1960, during my first summer of Ohio fieldwork, I visited a small, isolated tamarack bog a few miles from Kent. Working alone that day, I entered the bog and there, lo and behold, spread out before me were about twenty Labrador-tea shrubs. Here in Portage County, far from Greenland and Labrador, was a group of plants from the distant north offering a view of primeval Ohio few had seen. It was a day for a floristic botanist to remember.

THE FORMATION OF OHIO COUNTIES AND THEIR USE IN FLORISTICS

In Ohio, as in most other states, floristic fieldwork is often done within the confines of a single county or within a group of contiguous counties. The county's size and precise boundaries make it a useful geographic unit for such research.

While the settlers were clearing the forest and botanists were documenting the flora, Ohio was evolving politically. The Ohio region was originally part of the Northwest Territory, established by the U.S. Congress in 1788.[2] The first territorial governor, Arthur St. Clair, had served as a brigadier general during the Revolutionary War and was an admirer of his former commanding officer, George Washington, and of Washington's aide-de-camp, Alexander Hamilton. St. Clair soon established two large counties in the Ohio region, naming them Washington and Hamilton, respectively.[3]

Two weeks after he took office in 1788, St. Clair laid out Washington County, which included most of what is now the eastern half of Ohio. Hamilton County, which he created in 1790, had at first included only the land in the immediate vicinity of Cincinnati, but in 1792, St. Clair extended the boundary northward so that Hamilton County included most of what is now the western half of Ohio. Over the next two decades, these two huge counties were repeatedly subdivided. At the time of statehood in 1803, Ohio had nine counties. Among them were Washington and Hamilton, now greatly reduced in size. Another of the nine was Trumbull County in northern Ohio, established by St. Clair in 1800 and named for Jonathan Trumbull, the governor of Connecticut. It included all of what had once been the Connecticut Western Reserve.[4]

In 1808 Portage County, where Kent is located, was split off from Trumbull County, and Ravenna was designated as its county seat. Portage County today, however, is only two-thirds its original size. In 1840 the western third was used to form Summit County, along with smaller contributions from Medina and Stark counties.

The formation of new counties came to an end in 1851 with the establishment of Noble County, completing the state's roster of eighty-eight. Thus, by 1851 the stage was set for plotting Ohio plant distribution in terms of counties. Although varying somewhat in size, the counties have proved useful units for mapping plant distribution and have led to an interest among field botanists in discovering and collecting new county records.

This approach came prominently to the fore in 1950 when University of Cincinnati botanist E. Lucy Braun organized the Ohio Flora Project.[5] A major goal of the project was to improve knowledge of plant distribution within Ohio through the collection of new county records. At that time, such information was spotty. A 1961 study estimated that while the known flora of eight counties exceeded 1,000 species of vascular plants, for another fifty-eight counties (two-thirds of the total) it was fewer than 500 species.[6] And for a few counties, the known flora was fewer than 100 species. By the end of the century, new county records collected for Braun's Ohio Flora Project numbered in the thousands and the internal distribution of Ohio's plants was far better known than in 1950.

CATALOGS OF THE OHIO FLORA

If the discovery of a new county record is a satisfying event, even more gratifying for field botanists is the discovery of new state records. By the mid-nineteenth century, enough state records had accumulated to warrant compilation of the first official list of Ohio's plants.

In 1860 John S. Newberry published *Catalogue of the Flowering Plants and Ferns of Ohio,* which listed a vascular plant flora total of 1,377 species, 1,276 species regarded as native and 101 thought to be alien.[7] A series of such catalogs followed, appearing in 1874, 1893, 1899, 1914, 1932, and 2001, each accounting for the known Ohio flora of its day.[8] The 2001 catalog listed 2,716 species of vascular plants, 1,785 native and 931 alien.[9] The increase in the number of native species from 1860 to 2001 was the result of continued fieldwork in the state, while the increase in the number of aliens was largely due to a steady influx of foreign species into Ohio during that period.

CONSERVATION AND PROTECTION OF THE FLORA

In the 1970s the Ohio legislature enacted two significant statutes that affected the state's botany. The Ohio Natural Areas Act of 1970 made provision for the state to acquire and preserve important natural areas, and the Ohio Endangered Plant Law of 1978 provided protection for rare species of native Ohio plants. Both were administered by the Division of Natural Areas and Preserves (DNAP) of the ODNR.[10]

Several nongovernmental organizations have also been a part of the overall conservation and protection effort. The Ohio chapter of the Nature Conservancy, a private organization with a strong conservation agenda, was instrumental in the passage of both of the 1970s laws and continues today its work of saving important natural areas and of establishing nature preserves. The Ohio Biological Survey has published floras of special regions in the state and, in 1982, a book titled *Endangered and Threatened Plants of Ohio.*[11] Several Ohio arboreta, botanical gardens, and museums also conserve important natural areas and study techniques for preserving endangered plant species in cultivation.

As part of the Ohio Endangered Plant Law, the DNAP maintains an official list, updated biennially in even-numbered years, of Ohio's rare native plants. The 2008 list identifies some 600 species of native vascular plants (one-third of the 1,800 total) whose survival in the state is in some degree of jeopardy.[12] Each is assigned to one of four categories: potentially threatened, threatened, endangered, or presumed extirpated.

Nearly 100 species (5% of the 1,800 total) have not been seen in Ohio for the past twenty years and fall into the "presumed extirpated" category. These rarest species are not termed "extinct" because they are known to survive in other parts of their range, outside Ohio. If and when new Ohio populations are discovered, as has occurred in several cases in recent years, the species will be reassigned to one of the other, known-extant categories: endangered, threatened, or potentially threatened.

The DNAP has also published a *Directory of Ohio's State Nature Preserves,* with descriptions of each of the state's 130 dedicated preserves.[13] Some of these are owned and managed by the ODNR, some by the Nature Conservancy, and some by other organizations. These preserves, each protecting a segment of Ohio's primeval plant life, provide a connection to the past and a legacy for the future.

STATE WILDFLOWER

Ohio's state flower is the attractive, but nonnative, scarlet carnation. With increased interest in Ohio's indigenous plants during the rise of the environmental movement, the Ohio legislature selected a native flower as a second floral symbol for the state.

At 1:32 P.M. on the afternoon of December 4, 1986, Ohio governor Richard F. Celeste signed House Bill 763, which stated, "Be it enacted by the General Assembly of the State of Ohio: That section 5.021 of the Revised Code be enacted to read as follows: THE PLANT TRILLIUM

Large white trillium, state wildflower of Ohio, growing in a wooded area in Kent (Photograph by Gerry Simon)

GRANDIFLORUM, COMMONLY KNOWN AS THE LARGE WHITE TRILLIUM, FOUND IN EVERY OHIO COUNTY, IS HEREBY ADOPTED AS THE STATE WILD FLOWER."

Large white trillium (*Trillium grandiflorum*) is one of the largest and most often seen of all Ohio's wildflowers. It occurs in many local woods in the vicinity of Kent and can be found growing wild in a few places within the city itself. At Eagle Creek State Nature Preserve, near Garrettsville in Portage County, visitors can see a host of large white trilliums blooming each year in May. The flowers and green trillium leaves make a springtime carpet beneath the dark trees.

The new state wildflower has come to symbolize the floristic accomplishments of the period: an increased knowledge of the species comprising the Ohio flora, a great increase in knowledge of plant distribution within the state, new laws protecting rare and endangered members of the flora, and a statewide system of nature preserves. All these accomplishments were goals of the modern environmental movement.

BOWLING GREEN STATE
UNIVERSITY LIBRARIES

CHAPTER 3

THE HERBARIUM

A leaf pressed and dried between the pages of a book will retain its shape and, although fragile, will last indefinitely. This is the same principle by which scientific herbarium specimens are made.

In preparing an herbarium specimen, plant material is spread flat and arranged so as to show the species' diagnostic features, those features needed for accurate identification. The plant material is then dried in a special device called a press. After that, the specimen is glued to heavy paper, and care is taken to position it so that the important diagnostic features are visible. Also glued to the paper is a label with the name of the plant, the name of the collector, the location and date of the collection, and usually a statement of the plant's habitat. The resulting specimen is a basic tool of floristic research. A collection of such specimens is called an herbarium, and herbaria provide an understanding of the earth's flora.

For researchers in floristics and plant conservation biology, the herbarium specimen provides three basic pieces of data, namely that a particular plant was growing in a particular place at a particular time. When these data are gathered from thousands of specimens and considered together, they tell the composition of the flora of a county, a state, a geographic region, or a continent. For rare and endangered species, herbarium specimens provide essential information on the location of living populations or of places where the species occurred in the past. How important are the specimens? In *The Woody Plants of Ohio,* floristic botanist E. Lucy Braun answered the question succinctly: "Herbarium specimens are the only acceptable records of the occurrence of species."[1]

Because of their greater number of species and simpler body forms, vascular plants are often more difficult to identify than are vertebrate animals. Printed descriptions and illustrations may help in identification of unknowns, but in cases where two or more closely related species are much alike in appearance, it is often the technical characters visible on herbarium specimens that make accurate identification possible. Accurate identification is essential in all forms of botanical research, and for plant conservation work in which a large number of species are being considered, ready access to a local herbarium is invaluable.

Personal Background

For botanists working in the northeastern and midwestern United States, two masterful books published at midcentury prepared the way for the environmental movement that would follow. In 1950 Merritt L. Fernald's *Gray's Manual of Botany,* 8th edition, provided a much-needed, up-to-date "Handbook of the Flowering Plants and Ferns of the Central and Northeastern United States and Adjacent Canada."[2] In the same year, G. Ledyard Stebbins's *Variation and Evolution in Plants* introduced concepts of plant species in a book that became the main testament in the new field of plant biosystematics.[3] During the 1950s my wife, Miwako ("Mix"), a native of Hilo, Hawaii, and I were graduate students in botany at the University of Iowa. These books played an important role in the instruction we received and in our own research.

Mix and I had the same adviser, Robert F. Thorne. For his doctoral dissertation at Cornell University, Thorne had worked on a floristic research problem in Georgia, directed by Walter C. Muenscher.[4] (Muenscher had been a student of Charles E. Bessey at the University of Nebraska; Bessey was a student of Asa Gray at Harvard University; and Asa Gray, our academic great-great-grandfather, was the author of the first five editions of *Gray's Manual of Botany.*)

I was one of several of Thorne's graduate students who worked on the flora of Iowa.[5] In 1959 I published a floristic book on Iowa ferns with text, illustrations, and county dot-distribution maps.[6] *Gray's Manual* was my companion during the research for this book. Many years later, in 1988, I reviewed the history of the eight editions of *Gray's Manual* in a paper honoring Gray on the centenary of his death.[7]

Mix worked on a special problem involving introgressive hybridization, the movement of genetic material from one species to another following an initial cross between the two. The concept was then relatively new to the science, having been introduced by Edgar Anderson in 1949.[8] In 1957 Mix published a seminal paper on introgressive hybridization between two species of southeastern Iowa oaks.[9] Stebbins's book provided the context for her research. In 1993 Stebbins cited her paper on Iowa oaks in a retrospective summary of the most significant papers from the first sixty years of plant biosystematic studies in North America.[10]

Building a Regional Herbarium

During the 1960s, three Ohio botanists made major contributions to the study of the Ohio flora and its conservation. In 1961 E. Lucy Braun published *The Woody Plants of Ohio,* with text, illustrations, and county dot-distribution maps.[11] In 1962 J. Arthur Herrick circulated a draft of his survey of natural areas in Ohio. The draft, used widely as a guide by Ohio conservationists in the following years, was published in 1974 with the title *The Natural Areas Project.*[12] In 1966, using an ingenious plan of research, Robert B. Gordon created and published a map showing the vegetation of Ohio as it had existed in the early 1800s when the first European settlers arrived.[13] And, in 1969, Gordon released a book related to the map, *The Natural Vegetation of Ohio in Pioneer Days.*[14]

My first Ohio paper, in 1961, identified those counties of the state where the least amount of floristic work had been done and that were therefore most in need of attention.[15] This work was in preparation for my own later research on the Ohio flora and for that of future graduate students, research that would require a regional herbarium.

THE KENT STATE UNIVERSITY HERBARIUM

When I joined the Kent State faculty in 1958, I was put in charge of the herbarium and of overseeing its growth in support of the Ohio Flora Project. My new colleague J. Arthur Herrick served as a member of the project's steering committee. At that time, the biology department maintained a small collection of specimens curated by faculty member Clinton H. Hobbs. Earlier, Dr. Hobbs had registered the Kent State collection, assigned the abbreviation KE, with the editors of *Index Herbariorum,* an international directory of the world's herbaria. The 1959 index listed KE's founding date as 1921 and its size as about 1,500 specimens.[16] A few years later, I was formally named herbarium curator, and Mix was appointed herbarium assistant. In reality, Mix served as the director, handling the herbarium's daily operations and supervising its growth and development.

Mix was trained in herbarium work at the University of Iowa by Dr. Thorne, who brought to Iowa the procedures he had learned in the herbarium at Cornell University. She applied this heritage to developing the Kent State University (KSU) Herbarium. Over the next several decades, she trained scores of students, both graduates and undergraduates, in the exacting work of processing and mounting fragile plant specimens. She taught them also the procedures of professional herbarium management.

The first specimens added to KE were my collections from eastern Iowa[17] and western Virginia.[18] Supporting specimens from other nearby states were added as the herbarium grew. These specimens from surrounding areas often helped identify unknowns collected in Ohio. They also called attention to species not yet found, but to be expected, in Ohio and suggested the part of the state and the habitat in which they were most likely to occur.

During 1960 and 1961, I collected extensively from northeastern Ohio to provide future graduate students with reference specimens. The specimens included a number of new county records and two new state records, Louisiana sedge (*Carex louisianica*) and northern marsh bedstraw (*Galium palustre*).

Meanwhile, my first two graduate students were in the field collecting specimens for their research. (Both were awarded master's degrees

Tom and Miwako Cooperrider with KSU Herbarium's 50,000th specimen, May 1989 (Photograph by Ray Black)

in 1961.) Their specimens were accessioned into KE, and the pattern was set: new graduate students, more Ohio specimens, and an increasingly valuable herbarium. In all, twenty-two students working in that program under my direction contributed to knowledge of the Ohio flora while earning graduate degrees, the last awarded in 1996. New specimens also came in from undergraduate research projects and from the plant taxonomy classes' spring collecting trips, which were made to a different county in southern Ohio each year.

Purchases and gifts added several thousand specimens to KE.[19] An important purchase was the private herbarium assembled by Trumbull County botanist Almon Rood. Dr. Hobbs and I later wrote a sketch of Rood's life and his herbarium.[20] Allison Cusick donated a few thousand specimens from his collections in the eastern United States. Staff members of the ODNR and of the Ohio Chapter of the Nature Conservancy (TNC) donated more than 3,000 collections from Ohio nature preserves. These specimens, of unusual value, resulted from Mix's and my involvement in the Ohio Heritage Program from the time it began in the late 1970s.[21] The Heritage Program had been initiated jointly by the ODNR and TNC.

THE DAVEY SPECIMENS

Two gifts of local historical interest were the specimens made by Martin L. Davey in 1899 and those by Davey's future wife, Berenice Chrisman, in 1903. Both were class projects while the two were students at Kent High School. Davey was the son of John Davey, founder of the Davey Tree Expert Company, an international enterprise headquartered today in the city of Kent. Martin L. Davey later served several terms in the U.S. House of Representatives. He was elected governor of Ohio in 1934 and served two 2-year terms, during which Berenice served as first lady of Ohio. Most of Davey's collections were of cultivated plants. One of his specimens was a whimsical construction with his initials, MLD, made from small, dried flowers. Berenice's specimens were spring flowering plants from local woodlands.

The specimens were donated to KE by their daughter, Evangeline Davey (Mrs. Alexander) Smith. The Davey specimens were placed in specially made archival containers constructed by John Cooperrider.

THE EFFECT OF THE ENVIRONMENTAL MOVEMENT

The first celebration of Earth Day, on April 22, 1970, brought widespread recognition of the modern environmental movement that had been building in the United States for several years. The day was set aside to call attention to the earth's environmental problems. As the movement gained momentum, the focus changed from simply raising awareness to taking action.

Suddenly, the floristic work we had been doing primarily for its intellectual value took on added meaning. It became an important part of the new program of environmental action. Specifically, our work became part of a larger effort to conserve the plant life of the planet, especially its rare plants, and thereby maintain and improve the quality of human life.

In 1974 a special report, prepared for the National Science Foundation, evaluated the nation's herbarium resources.[22] A statement in the report's introduction noted that "the nation and the scientific community have awakened to the pressing need to have more and better information about the plants and creatures of the earth's biosphere."[23] For plants, the prime source of more and better information was to be found in the floristic collections of the nation's herbaria.

In the report, some 1,100 U.S. herbaria were ranked on the basis of size, quality, facilities, and staff, and the top 105 were designated "National Resource Collections." In this group were the herbaria one might expect to find, for example those of the Ivy League universities, the Big Ten schools, and the Pacific-10 conference. The top group also included the herbarium at Kent State University.[24] In sixteen years, and mainly as a result of Mix's diligence and dedication, KE had grown from a small collection of some 1,500 specimens into one with a place among the nation's finest. The Kent State herbarium was probably the smallest in the top group, but KE was there nevertheless, and in good company. At the other end of the scale was the largest university collection, Harvard's five million specimens.

Building the KSU herbarium, especially at a time of increased attention to the earth's future, was gratifying work. Among other environmental applications, KE played a major role in assembling new data on Ohio's native flora and in compiling Ohio's first comprehensive list of rare native plants. In May 1989 we accessioned KE's 50,000th specimen.[25]

CHAPTER 4

A COLLEAGUE

A colleague who shares one's goals and whose work complements one's own is like a comrade in arms. For me, such a colleague was James Arthur Herrick. While Art, as he was known, inventoried the natural areas of Ohio, the places where the state's rarest native plants grow, I studied the Ohio flora, for which the natural areas were a major source of information.

Art Herrick began his inventory at a time when attitudes about protecting and conserving natural areas were changing. He visited naturalists throughout Ohio, and if they had secret haunts—such as a woodland with old-growth trees, an isolated prairie patch that had never been plowed, or a wetland that had never been drained—he persuaded them to add these favorite sites to his inventory. Like Art, they saw that the pace of destruction and development had increased rapidly and that saving their prized natural areas depended on making knowledge of them public.

A native of Twinsburg in Summit County, there was a strong element of New England Yankee in Art's makeup. An early Herrick ancestor immigrated to Connecticut from England in 1630. After the family had been established there for several generations, a more recent ancestor joined many other pioneers in the move west to New Connecticut, also called the Connecticut Western Reserve, an area that encompassed much of what is now northeastern and northcentral Ohio. A little larger than Connecticut, it was for a time somewhat like a seventh New England state. After Ohio's admission to the Union in 1803, the area became known simply as the Western Reserve, but its New England traditions persisted.

In the late nineteenth century, residents of rural townships in the Western Reserve prided themselves on living in an area "more thoroughly New England in character and spirit than most . . . towns in New England today"—according to a famous Western Reserve son, President James A. Garfield.[1] It is not hard to believe that this heritage figured into Art Herrick's work habits. His annotated inventory of Ohio's surviving natural areas calls to mind a New Englander's end-of-the-season accounting.

Art Herrick spent most of his adult life in the city of Kent. He was a faculty member at Kent State University from 1938 to 1972, with a leave of absence to teach at the University of Michigan medical school from 1943 to 1946. After returning from Michigan, he lived in Kent for the next sixty years.

The Work of a Twentieth-Century Kent Botanist

Art Herrick grew up working with plants. The family orchard, located in Twinsburg, Ohio, was a sizable operation. It included fifty acres of apple trees and another twenty acres of peaches, cherries, plums, and berries. The orchard required enough labor to keep all six children of the family busy. The four oldest were boys, the two youngest girls. Art was the fourth child, the youngest son.

He learned how to trim the orchard's trees and shrubs when he was young. Good trimming of woody plants is partly an art, partly a science. It was a skill that Art Herrick practiced throughout his life, one that gave him great satisfaction and pleasure.

Art once told me of an incident from his childhood that throws light on his character and outlook on life. The story concerned his grandfather, who raised sheep and whom Art described as "a frugal man." I know something of sheep from my own childhood on our family farm. An individual sheep would sometimes wander off and die or get killed, and, if part of a flock, its absence might go unnoticed for several days. In such cases, the value of the animal was lost, but, unless it had recently

Biology Department Faculty, Kent State University, 1959. Front row: Harry Cunningham, Dorcas Anderson, Ruth Kelley, Charles Riley; 2nd row: Charles Sumner, Emanuel Hertzler, Vincent Gallicchio, Ralph Dexter, Clinton Hobbs, Peter Zucchero; 3rd row: J. Arthur Herrick, Kenneth Kelley, Tom Cooperrider, George Easterling, Adam Cibula. All the faculty members shown here with Dr. Cunningham were hired during his tenure as department head, 1927–1959. (Photograph Al Schleider)

been shorn, it had a coat of wool that could be harvested and sold for good money. After the body begins to decompose, the wool comes off easily, but still someone has the task of collecting and bagging the wool. His grandfather would pay Art and a cousin 25 cents for collecting the wool from a dead sheep. Bag in hand, the boys would set off to find the carcass. On the way, they stopped to collect a few mint leaves to stuff in their nostrils. At the end of the day, they had earned some money for performing an unpleasant task, experienced their grandfather's example of thrift and avoidance of waste, and, incidentally, learned the value of teamwork and of being able to identify wild mint plants.

EARLY YEARS

I met Art in the spring of 1958. He was a professor in the biology department at Kent State University. My wife and I had driven from Iowa City to Kent, where I was interviewing for a position in floristic botany, and Art was our host. I later accepted the job offer and joined the department in September, gaining a set of colleagues whom I soon came to admire and whose interests ranged across the broad spectrum of biology. Art was fifty years old at the time and about to start on the great conservation adventure of his life, a survey of Ohio's natural areas.[2]

Years later, Art gave me reprints of papers he had coauthored in the early 1940s while he was an instructor in the medical school at the University of Michigan. One paper concerned the effect of penicillin on bone infection in rats.[3] A handwritten note from Art attached to the reprint read: "This was the first bone infection cured by antibiotics." The work was done during World War II when penicillin was new on the medical scene and had already proved useful for military doctors. But despite this interesting and valuable research, Art's calling was in another area. He left Michigan in 1946 and returned to Ohio and Kent.

NATURAL AREAS

Art was a native Ohioan. He loved nature, and he believed in conservation. These three elements of his life all came together in 1958 when, in December, he began work on the Natural Areas Project sponsored by the Ohio Biological Survey. The purpose of the project was to "assemble and evaluate data on natural areas of interest to biologists, naturalists, teachers and conservationists."[4] Its goal was to make the first comprehensive inventory of all the natural areas remaining in the state. This was about 150 years after European settlers had begun to clear the forests and to make the region ready for farming and other activities of civilized society. Each of the surviving natural areas provides a glimpse of primeval Ohio, and each natural area is a stronghold of native Ohio plants.

Art spread the word far and wide. He loved talking to people about the work, and he did this in an engaging manner that aroused their interest and enlisted their support. As he wrote in the introduction to his 1974 book *The Natural Areas Project: A Summary of Data to Date,* "During the early years of this project I personally addressed many groups throughout Ohio to explain our aim and our methods. Forms for reporting areas were made available. As reports came in and time allowed, I personally inspected many areas. In a majority of cases, I did the inspection with the guidance of the local naturalists who had sent in the original report."[5]

Another of Art's virtues, his love of people, came into play during this project. "On many inspection trips," he wrote, "I enjoyed the hospitality, food, lodging, transportation, and guide service of local naturalists."[6] When the project ended, Art had assembled data on nearly 600 natural areas in the state and had inspected some 200 of them. By then he was known, and admired, by nearly every naturalist in Ohio. The Natural Areas Project was a personal triumph for Art. It came at a time when interest in conservation was increasing throughout Ohio and across the nation.

His 1974 publication, like the widely distributed mimeographed drafts that had preceded it in 1962 and 1965, came to be called simply Herrick's List. Cataloguing natural areas from eighty-six of the state's eighty-eight counties, it was used to guide the work of the Ohio chapter of TNC and that of the ODNR Division of Natural Areas and Preserves, as well as other Ohio organizations, in their establishment of protected nature preserves throughout the state. Art became directly involved in securing some of these natural areas, notably Herrick Fen Nature Preserve in Portage County, a few miles north of Kent. Herrick's List is still used today.

SIMPLE GIFTS

Art liked giving people young trees to plant around their homes. Around his own home, he had developed an arboretum of about 1.5 acres. In it were 400 species of woody plants he had collected during his travels in the eastern United States. As these plants matured and shed seeds, he would mark their seedlings for protection with a stick placed in the ground beside each one. When the seedlings were older he moved them to a better site, or to a container, and nursed the seedlings into saplings. They made gifts that he gave to anyone who wanted them. With the least encouragement, Art would deliver the plant, select the best location, and put it in the ground.

We have two of Art's gifts in our yard. One is a fringe-tree (*Chionanthus virginicus*) grown from seeds derived from southern Ohio trees. It blooms profusely each spring, filling the evening air with a sweet fragrance. On one occasion I told Art I was looking for a Fraser fir (*Abies fraseri*), a tree native to the Appalachians, to plant on the slope behind our home. He immediately said, "I have one I'll give you." I envisioned a young tree, at least one foot or more in height. Art lived around the corner, about a block away, and a few days later he showed up at the door carrying an old, crumbling basket filled with soil. Looking carefully, I could see that there was a speck of green in the center. It was a Fraser fir seedling germinated in the spring and was less than a half-inch tall. With a slight twinkle in his eye, Art said, "Just plant the basket and all." I did just that. Now, several years later, the seedling has grown into a tree that is more than three feet in height. I have lately seen birds sitting on its low branches.

Art also gave us a couple of his handmade tree-of-heaven walking sticks. Tree-of-heaven (*Ailanthus altissima*) is a weedy tree of Asian origin naturalized in disturbed places throughout the Kent area. The tree produces smooth sturdy basal shoots that can grow more than ten feet in a single summer and attain a diameter of one inch or more. At the end of summer, Art would cut from the shoot a straight section of walking-stick length. After a little seasoning, the stick was ready for use. This was the summer model. The winter model had an additional feature. In one end of the stick, Art would pound a nail, allowing it to protrude about an inch. He then cut off the nailhead, leaving the end sharp and sometimes a little jagged, making the stick useful when walking on ice. We have our walking sticks outside the back door and use them when climbing the hill where the Fraser fir grows. That they were made by hand, with essentially no expense, pleased Art, and us, very well.

PARTING GIFT

Art left Ohioans a priceless, original gift, his annotated inventory of the state's natural areas. At the age of sixty-six, thinking back on the effort that had gone into the collection of the basic data and the composition of the report, he wrote, "The past fifteen years have been productive, pleasant and very rewarding in terms of personal satisfaction, as this work has been a major part of my way of life."[7] He continued working for natural area conservation for the remainder of his years.

Art Herrick lived to be a centenarian. He was born on July 5, 1908, and he died on July 20, 2008.

CHAPTER 5

THE CAMPUS PLANTS AND GARDENS

Cultivated plants play a prominent role on an educational campus such as Kent State's. From jonquils in the spring and chrysanthemums in the fall to the stately old oak trees, the campus plants have an aesthetic as well as an instructional value.

The Kent State campus also has an unusual use of plants as a commemorative. Spread across a wooded hillside is the botanical memorial, *58,175 Daffodils,* honoring the American military personnel who lost their lives in the Vietnam War. The artistic idea of Brinsley Tyrrell, the daffodils were planted by the grounds department in the late fall of 1989. They bloomed the following spring in time for the twentieth anniversary of May 4, 1970, the day on which four students were killed and several wounded during an antiwar protest on the KSU campus. Since 1990 the golden flowers have bloomed each spring, brightening the slope below another commemorative, the May 4 Memorial, dedicated to the slain students. A ground-level metal plaque identifies the daffodil site.

From 1910 to 1949, some fifteen buildings were constructed on the school's campus, including classroom, dormitory, and service buildings. Trees, shrubs, and wall-climbing vines were the principal cultivated plants on the original campus. Many of those plants survive today, along with a few large, white oaks left from the farmstead on which the campus was built in 1910. Among the cultivated survivors are a wide variety of hardwood trees planted by Larry Wooddell, longtime superintendent of grounds. Another survivor on the older part of campus, and still a favorite place in spring, is Lilac Lane, a walkway lined on both sides with many cultivated varieties of lilacs. A large boulder with an identifying plaque is located at the entrance to the walk just east of Oscar Ritchie Hall.

During the postwar boom decades of the 1950s and 1960s, some fifty new buildings were constructed as the campus grew and expanded eastward. The bleakness of the recently cleared land around the new buildings was relieved by the planting of gardens of ornamental plants.

Four gifted horticulturalists led the next phase of transformation and growth of the campus landscape. Three of the four were managers of the greenhouse attached to Cunningham Hall, which housed the Department of Biological Sciences. I was a member of the biology faculty at that time, and from 1968 to 1993 I served as the liaison faculty member for the departmental greenhouse managers. Although overseeing the greenhouse was their main task, each manager found some time to work on creating gardens near Cunningham Hall.

Ruth Rogers, the first greenhouse manager, served from 1968 to 1973 and established a program that provided plants for class experimentation and dissection and began a conservatory of tropical plants for instructional use. Outside the greenhouse, she planted the first gardens, a few beds of flowering plants. In the days immediately following the protests on May 4, 1970, Ruth was one of the few people allowed on campus. She never forgot the eerie sensation of entering through a military checkpoint and walking across the quiet, deserted campus to water the plants in the greenhouse. Ruth went from Kent State to the New York Botanical Garden, where in 1989 she published her masterwork, *Perennials for American Gardens.*[1]

Jayne Timmerman managed the greenhouse from 1973 to 1979. She and her student helper, Tana Smith, designed and built the Alumni Garden, aided by local nurseryman Ernie Miller. In a secluded area east of the greenhouse, the garden was comprised of several plots of perennials, shrubs, and small trees. A chemistry professor who passed the garden daily in the 1970s called it, at that time aptly, "the most beautiful unknown site on the Kent Campus." It also functioned as an outdoor classroom.

The plots were later named the C. V. Riley Alumni Garden, in honor of Dr. Charles V. Riley, former biology department chair.

Christopher Rizzo began managing the greenhouse in 1979. Outdoors, he and I established the Magnolia Gardens (described later).

Michael Norman joined the KSU staff in 1978 as a horticulturalist with the grounds department. I did not work with Mike, but I knew him as a friend and also as a gardener of great creative talent. He turned his attention first to the Judith A. Beyer Murin Memorial Gardens, a small group of plantings set out in 1975 near the library and the student center.[2] There he added an array of showy ornamentals, enlarging the garden to about five times its original size. Prominently located, the Murin Gardens became an immediate campus showpiece. He built a number of gardens in the following years, and by the time of his retirement in 2002, his splendid and varied gardens had become signature pieces of the KSU campus.

Paralleling the earlier award to the city of Kent, in 2009 the Arbor Day Foundation designated the Kent State campus a "Tree Campus USA," the first such designation in Ohio, and one of only a few in the nation.[3] Heather White, manager of campus environment and operations, directed the application effort for the award.

The Herrick Magnolia Gardens

In the early 1980s, Christopher Rizzo and I designed and built the Magnolia Gardens at Kent State University. Chris served the Department of Biological Sciences as director of horticultural facilities, as he still does today. I served as the departmental director of botanical gardens.

The Magnolia Gardens include some fifty trees and shrubs. A few are located north of Cunningham Hall, but the majority are in plots north of Henderson Hall, the nursing building, adjacent to Cunningham. Some of the plants were donated by Dr. J. Arthur Herrick from the arboretum he maintained at his home in Kent, some were provided by the KSU grounds department from their holdings, and some I collected from my home garden or from the wild. Other plants were purchased from a variety of sources.

Magnolia Gardens north of Henderson Hall, Kent State University campus. Japanese magnolias in bloom in mid-April (Photograph by Bob Christy)

University administrators Lowell Croskey and Chester Williams secured approval for the project, including for the location of the site. Because the area is low, drainage pipes were laid in parts of it. In addition, we built substantial flat-topped hills on which to plant the trees, elevating them above the water table. Russell Foldessey, manager of the grounds department, provided the heavy equipment and labor needed to lay the drainage pipes, construct the hills, and plant the larger trees. Chris Rizzo supervised all the planting and did much of it himself.

The plants are now about twenty-five years old and have reached flowering maturity. Most of the species bloom in the spring months of April and May, and most have fragrant flowers. The Ohio and other North American species in the Magnolia Gardens are described in *Flora of North America.*[4] The Asian species and hybrids are described in *Hortus Third.*[5]

NOMENCLATURE AND CLASSIFICATION

Plants have been given names in every human language. As our contemporary civilization evolved, plant species came to have two different kinds of names: a single scientific name in Latin used in formal science, such as *Magnolia acuminata,* and one or more common or vernacular names in the modern languages of everyday life, such as cucumber magnolia.

Some of the magnolias provide examples of a frequent phenomenon in plants, the occurrence of a hybrid between two species. A multiplication sign inserted in the middle of a scientific name, for example, *Magnolia* × *loebneri,* indicates that the plant is an interspecific hybrid resulting from the cross of two species, in this case *Magnolia kobus* × *Magnolia stellata.*

In order to deal with the half-million plant species in the earth's flora, botanists arrange them in groups, using a classification system with seven principal ranks. The ranks, in descending sequence, are kingdom, phylum, class, order, family, genus, and species. The number of ranks can be doubled by creating seven more with the prefix *sub-*, for example, subkingdom, subphylum, subclass, and so on down to subspecies. Each rank has a Latin scientific name.

The plants in the Magnolia Gardens all belong to a subclass with the scientific name Magnoliidae, an assemblage that includes magnolias and their close relatives. This is a group of unusual importance because they are thought to be among the most primitive of all living flowering plants. Most members are tropical, but those in the Kent State campus garden are hardy in the temperate zone.

NATIVE OHIO SPECIES

Today, in a small area, visitors can see all the woody members of the subclass Magnoliidae native to Ohio. These species, eight in all, are listed at the beginning of the *Seventh Catalog of the Vascular Plants of Ohio.*[6]

Heading the list are the four Ohio species of the magnolia family. All are deciduous trees, that is, ones that shed their leaves each fall. Cucumber magnolia (*Magnolia acuminata*), also called cucumber-tree, is native throughout eastern Ohio and is the only true magnolia native to the Kent area. The large tree has fragrant, greenish-yellow flowers. The tree's common name comes from its fruit, which often fall prematurely. They are frequently seen on the ground beneath a tree in early summer, resembling a scattering of small green cucumbers. Bigleaf magnolia (*Magnolia macrophylla*), true to its name, has leaves three feet long. Native in extreme southern Ohio, it has large, fragrant, white flowers. Umbrella magnolia (*Magnolia tripetala*), native to southcentral Ohio, has white flowers that are somewhat ill-scented. Tulip-tree (*Liriodendron tulipifera*), also called tulip-poplar or yellow-poplar, is native in the Kent area and throughout most of Ohio. It has large, lightly fragrant, yellowish-green flowers reminiscent in form and size of cup-shaped tulips. The other four species—pawpaw, sweet-shrub or Carolina-allspice, spicebush, and sassafras—all belong to other families closely related to the magnolias.

Using the heavy equipment available, we moved three mature pawpaw trees (*Asimina triloba*) onto the site and planted them in a group. Pawpaw, a member of the custard-apple family, has brownish-purple flowers with a slightly fetid odor. The edible, gamy-tasting fruit is highly prized by some. Pawpaw is native throughout Ohio but is more frequent in the southern counties. The three initial trees have given rise to a number of saplings, producing all in all a "pawpaw patch" like those found in southern Ohio.

Sweet-shrub or Carolina-allspice (*Calycanthus floridus*), native in southeastern Ohio near the Ohio River, is the state's sole representative of the strawberry-shrub family. Its flowers are reddish-brown and have an unusual fragrance, variously described as being similar to the odor of cantaloupe, pineapple, or strawberry.

The plantings in the Magnolia Gardens also include two local species of the laurel family. One, spicebush (*Lindera benzoin*), is a shrub, and the other, sassafras (*Sassafras albidum*), is a small- to medium-sized tree. Both have small, fragrant flowers. Both are native to the Kent area and throughout most of Ohio.

MAGNOLIAS OF THE SOUTHEAST

The Magnolia Gardens also include two species of magnolias native to the southeastern United States but whose ranges do not extend as far north as Ohio. There are several plantings of sweet-bay (*Magnolia virginiana*), a small tree with semi-evergreen leaves and very fragrant white flowers. It blooms from late spring into early summer and often again from late summer into autumn. The gardens also include a single specimen tree of the

Southern magnolia, leaves and flower bud, outdoors near Herrick Conservatory, Kent State University campus (Photograph by Gerry Simon)

Southern magnolia in flower, outdoors near Herrick Conservatory, north of Cunningham Hall, Kent State University campus (Photograph by Gerry Simon)

famous southern magnolia (*Magnolia grandiflora*). The tree, with its large, glossy, evergreen leaves, grows outdoors in a protected site near the Herrick Conservatory, north of Cunningham Hall. Although far from its home territory, the southern magnolia tree blooms regularly each summer. Its flowers have large, creamy, white petals and a strong, lemony fragrance.

ASIAN MAGNOLIAS

Visitors to the gardens can also see living representatives of several Asian magnolias, species and hybrids, from the North Temperate Zone in China and Japan. All are small deciduous trees.

The Asian trees' floral display varies from year to year. A few untimely warm days in early spring followed by a night with a hard frost can leave the trees with a large crop of dead, brown flowers. In years with a more favorable warming sequence, the flowering display can be spectacular.

Two of the Asian magnolias in the gardens have colorful, fragrant flowers that bloom in late spring. One is the Chinese species, lily-flowered magnolia (*Magnolia liliiflora*), which produces purple flowers. The other, displaying pink flowers, is a hybrid called Chinese magnolia or saucer magnolia (*Magnolia* × *soulangiana*). This hybrid results from a cross between lily-flowered magnolia and another Chinese species, the white-flowered Yulan magnolia (*Magnolia denudata*). The gardens have no Yulan magnolias.

The showpiece of the Magnolia Gardens is a group of white-flowered magnolias, two species and their hybrid, all of Japanese origin. On a low, broad hill are a few trees of Kobe magnolia (*Magnolia kobus*). Its fragrant flowers have—usually—six broad, white petals. On an adjacent hill are several small trees of star magnolia (*Magnolia stellata*). Its flowers, with numerous, narrow, spreading white petals, are scarcely to only slightly fragrant. Beside the two species are several trees of their hybrid (*Magnolia* × *loebneri*), with flowers intermediate between those of the parents.

The two Japanese species and their hybrid bloom at approximately the same time, usually about mid-April. Because the trees are still bare of leaves and their flowers are large, the floral display is striking. Viewed from the top (twelfth) floor of the nearby KSU Library, the area appears to be covered with a cloud of large white flowers.

DEDICATION AND SIGNIFICANCE

On September 17, 1996, a ceremony honored my longtime colleagues and friends, J. Arthur Herrick and Margaret Hatton Herrick, in appreciation of their benefactions to KSU. During the event the university dedicated in their honor two botanical features of the area, the conservatory behind Cunningham Hall and the magnolia garden on the north side of Henderson Hall. A plaque affixed to a large boulder behind Henderson Hall marks the magnolia garden area.

Located in the North Temperate Zone, the Herrick Magnolia Gardens at Kent State provide students and other visitors with a unique opportunity to see and study a collection of living relatives of the tropical plants thought to have been the ancestors of all flowering plants. It is the only such garden in Ohio.

CHAPTER 6

A SPECIMEN TREE

It is a common occurrence to see an American holly tree with a large group of holly seedlings growing beneath its lowest branches. If an attractive seedling is dug up and planted elsewhere, it immediately becomes a cultivated plant. This act, however, does not make the seedling a cultivar. That special status is achieved only when a plant is given a cultivar name.

The word *cultivar* was coined by combining the first part of two words: *cultivated* and *variety.* A cultivar is a cultivated variety of sufficient importance to be given a name. The authors of *Hortus Third* write, "Cultivar names are now formed from not more than three words in a modern language and are usually distinguished typographically by the use of single [rather than double] quotation marks."[1] Familiar examples are 'Better Boy' tomato, 'New Dawn' rose, and 'Heavenly Blue' morning glory.

The editors of *Hortus Third* note that American holly has more than one thousand named cultivars.[2] This extraordinarily high number reflects the popularity of this tree in cultivation and also the great number of available seedlings a single holly tree may produce. Each seedling has the potential to become a new holly cultivar. These are not, however, a thousand new species. All these cultivars belong to a single species, *Ilex opaca.*

One of the American holly cultivars is named 'Beautiful Ohio.' The search for this famous tree took me back to Dawes Arboretum in Licking County. In 1941, when I was a junior high school student in Jacksontown, a few miles away, I won a tree identification contest at Dawes Arboretum. My knowledge of trees came chiefly from those growing on our family farm, where I learned their names from my parents, both of whom were interested in plants. I am a Licking County native, and I was born in Newark, the county seat.

American Holly, George Washington, and 'Beautiful Ohio'

It was the kind of tree one stops to admire, an American holly described by Galle as having leaves of "dark olive green" and fruit of "deep reddish orange."[3] The year was 1959, and the tree was in the front yard of a house on Summit Street in Kent, about a block from McGilvrey Hall, where Kent State's biology department was located. New to the university the previous year, I spent much time surveying the plants in the vicinity of McGilvrey.

American holly 'Beautiful Ohio' at Dawes Arboretum (Photograph by Margaret Popovich)

'Beautiful Ohio' American holly tree at Dawes Arboretum, Licking County (Photograph by Margaret Popovich)

The owner, who came out to greet me, said the tree was a cultivar named 'Beautiful Ohio.' Being a native Ohioan and knowing the state song of the same name, I enjoyed the reference. After a brief conversation, I walked on, storing the information away for a few decades. Forty years later, I learned that the man with whom I had spoken was Joseph G. K. Miller. Mr. Miller worked at the Davey Tree Expert Company in Kent for forty-six years. He has now passed away, the house has changed hands once or twice, and the tree is gone.

When I prepared the section on the holly family for the Ohio Flora Project, I recalled the 'Beautiful Ohio' tree along Summit Street and mentioned it in the text.[4] The story might have ended there had I not learned by chance a few years ago that a Kent resident and good friend, Audrienne Galizio, whom I have known for many years, is Mr. Miller's daughter. She remembered the holly tree at her childhood home on Summit Street and how it came to be planted there.

As a part of his work for the Davey Tree Expert Company, Mr. Miller was in charge of tree restoration at Wakefield, the Washington family estate and George Washington's birthplace. Near the Potomac River in Virginia, Wakefield is about seventy-five miles downstream from Washington's more famous later home, Mount Vernon. In 1933, while working at Wakefield, Mr. Miller came across a good-looking seedling of American holly. He brought it back to Kent and planted it in his front yard. It grew into an attractive tree, and he and his Davey coworker M. W. "Biff" Staples decided to give it the cultivar name 'Beautiful Ohio.' Staples registered the name with the American Holly Society and deposited a supporting herbarium specimen taken from the tree at the U.S. National Arboretum.[5]

WILD HOLLIES

Although native in southernmost Ohio, American holly (*Ilex opaca*) is not native to the Kent area.[6] However, it does well in cultivation and is popular as a small ornamental tree because of its shiny evergreen leaves and red berries. Occasionally, an individual plant will be seen in the wild, usually in a weedy field, where it has escaped from cultivation.

Winterberry (*Ilex verticillata*), also called winterberry holly, is a native plant of moist places in the Kent area. A deciduous shrub, it is related to American holly but differs markedly in appearance. In early winter, after its leaves have fallen, the red holly berries are conspicuous on the shrub's naked branches. This species, like American holly, has been taken into cultivation, and it also has many named cultivars. A few wild, native winterberry shrubs grow along the boardwalk trail at the Kent Bog State Nature Preserve.

PROPAGATION AND SURVIVAL OF CULTIVARS

The most common way to propagate a holly cultivar is to plant a cutting, a small branch taken from the original tree. If the attempt is successful, the cutting will grow into a genetically identical clone of the parent. Today

Winterberry holly at Kent Bog State Nature Preserve (Photograph by Mix Cooperrider)

modern laboratory programs can also produce a clone, grown in a test tube, from one or a few cells taken from the parent plant. Seedlings, however, although similar to the parent tree are not genetically identical to it and cannot be given the cultivar name of the parent. I wondered, had Mr. Miller given cloned specimens of 'Beautiful Ohio' to any Ohio arboreta? After several inquiries I learned that he had.

With the help of Ken Cochran of Secrest Arboretum in Wayne County, Ethan Johnson of Holden Arboretum in Lake County, and Bonnie Beeman and Greg Payton of Dawes Arboretum in Licking County, I was able to locate two living 'Beautiful Ohio' holly trees. One is at Secrest Arboretum, the other at Dawes. The Secrest tree, marked with a metal tag, is in a plantation of some fifty holly trees set out in rows. The Dawes tree, the more vigorous of the two, is part of a group of a hundred or more widely spaced holly specimen trees and shrubs on Holly Hill. Both trees were planted in about 1970. There may also be living trees at other arboreta.

The waltz "Beautiful Ohio" lives on as Ohio's official state song. If the holly of that name is to be perpetuated by cloning, it must be done while these known trees are still living. An attractive tree with a name of special interest, it would seem appropriate to have this cultivar at each of Ohio's major arboreta as well as at other important sites in the state. The trees would be a lasting botanical gift from a tree that once grew in the city of Kent.

CHAPTER 7

THE BOG

Tamarack bogs occupy a unique place in the Ohio flora. Their plants are distinctive, most are not found in any other habitat, and nearly all the plants in the bogs are native, an unusual situation in any habitat in Ohio today.

Kent Bog State Nature Preserve is mostly located within the city of Kent, about two miles south of the downtown area. A small part of the preserve lies outside the Kent city limits in adjacent Brimfield Township. The bog itself extends a short distance beyond the boundaries of the preserve, onto privately owned land, also in Brimfield Township.

The bog's general location adds to its significance. It is thought to be the southernmost tamarack stand of this size in the United States, making it valuable to the nation as well as to the state.

Within a radius of fifteen miles are six other nature preserves with tamaracks. Together these seven preserves hold a large majority of the tamarack trees in the state. The city of Kent could rightly be called the tamarack capital of Ohio.

Among the other tamarack stands in the immediate area, an important one is that at Triangle Lake Bog State Nature Preserve, about five miles to the east of Kent. Like the Kent Bog, Triangle Lake Bog also is owned and managed by the ODNR. These preserves make an interesting pair for study. Both bogs are thought to be approximately the same age. But in the natural transition from lake to bog, Triangle Lake Bog is at an earlier stage of development than is Kent Bog, Triangle having at its center a small, acidic, glacial lake.

In the past, the bog at Kent probably looked much the same as the bog at Triangle Lake looks today, with a small lake encircled by a ring of tamarack trees. Presumably, the Kent lake was somewhat diamond shaped as opposed to triangular. The surface of the bog now occupying the basin is flat, like the surface of the lake that was once there.

Kent Bog State Nature Preserve

> You are about to enter a very special place where time has virtually stood still since the Ice Age.
>
> —Guy L. Denny, quoted on a plaque at the entrance to Kent Bog State Nature Preserve

Located in Portage County, in northeastern Ohio, Kent Bog can be likened to a large, natural container garden in a deep, clay bowl. The bowl, about forty feet deep, is filled with water-laden peat. All the openings in the peat substrate are filled with water. The garden, about forty-five acres in size, consists of tens of thousands of individual plants growing at the peat's surface.

At first the clay bowl, or basin, held only water, a small lake derived from glacial ice. In nineteenth-century Ohio, when large iron and copper kettles were a part of everyday life, these more or less circular glacial lakes came to be called kettle lakes or kettle hole lakes. The bogs, such as Kent Bog, that later formed in the sites were called kettle bogs or kettle hole bogs. These terms are still in use today.

Aerial photographs show that the bog is roughly diamond shaped. Along most of its periphery, and visible when on foot, are two usually well-differentiated zones. The outermost zone is a low bank, the rim of the basin. Growing on the bank are trees of a few hardwood species common in this region, mostly red maple, red oak, sour-gum, and wild black cherry. The next zone, on the inner side of the bank, is an irregular

Kent Bog, 1984 (ODNR/DNAP aerial photograph)

moat—also called by the Swedish term *lagg*—in which there is shallow standing water much of the year. Inside the moat is the bog.

THE MAJOR PLANTS

Across the entire bog, growing directly on the peat, is a thick carpet of sphagnum moss. Here and there are patches of a low-growing, rare plant, small cranberry (*Vaccinium oxycoccos*), and several species—some also rare—of slender grasslike sedges. A taller plant, Virginia chain fern (*Woodwardia virginica*), grows in extensive colonies throughout the bog.

Four main species of woody plants grow in the bog. Leather-leaf (*Chamaedaphne calyculata*), a low shrub, grows near the moat. Highbush blueberry (*Vaccinium corymbosum*), a taller shrub, occurs in dense stands throughout the bog. Gray birch trees (*Betula populifolia*) are scattered here and there, growing singly or in clumps of a few individuals. Tamarack trees

(*Larix laricina*), towering over the other plants, are mostly congregated in the large central core of the bog. With more than 2,000 individual plants, this is the largest stand of tamaracks in Ohio. It is also an actively reproducing population, with many seedlings and saplings.

Red maple (*Acer rubrum*), a native tree common to moist woodlands and to floodplains, is an invasive plant in the bog. Saplings and stunted trees of red maple are cut out from time to time to prevent their displacement of native bog plants.

A LARGE BLOCK OF ICE

Sixteen explanatory plaques have been placed along the loop trail that runs through the preserve. The first three are devoted to the bog's geological and botanical history. This is the story they tell.

At the end of the Ice Age, about 12,000 years ago, a large piece of ice became separated from the glacier that covered this region. It came to rest at a site just south of what is now Meloy Road in western Portage County. The huge ice block was quickly buried under enormous amounts of clay, silt, sand, and gravel being released from the melting glacier. The clay particles and other materials had become incorporated into the glacier as it scoured the bedrock on its southward journey into Ohio. As the buried ice block melted, these particles settled to the bottom, forming the clay basin for a kettle hole lake.

Boreal (northern) plants, such as tamarack and leather-leaf, that had moved south in front of the advancing glacier, now colonized the wet ground around the new lake. The climate of the entire area eventually warmed and the front of the glacier melted away. In most places during this warming period, the boreal plants died off, and other plants from warmer zones moved in to take their place. But in the bog that formed around the lake, the northern plants from colder lands survived.

As individual plants in the bog died, they did not decay completely. Fungi and bacteria, the principal decomposers, were unable to function effectively in the bog conditions. Little by little, the partially decomposed plant material, called peat, accumulated in and around the lake. Eventually the basin holding the lake was completely filled with peat, much of it derived from sphagnum moss.

The last vestige of an open lake at the Kent Bog site probably disappeared prior to European settlement. Meanwhile, the bog plants had moved inward from the lake's margin and taken up residence across the peat substrate, producing the kettle hole bog we see today. The plants now living in the bog are descendents of the boreal plants that came south thousands of years ago. A low bank that once surrounded the glacial lake now surrounds the bog.

BOTANICAL SPECIMENS

On July 30, 1961, I visited Kent Bog for the first time. I was accompanied by Thomas Zavortink, a KSU biology major from Ravenna. We entered the bog from the north, at the Stark family residence on Meloy Road, and went in only a short distance. I collected specimens of small cranberry (*Vaccinium oxycoccos*), three-seeded sedge (*Carex trisperma*), and tawny cotton-grass (*Eriophorum virginicum*). On the specimen labels, I described the habitat and location as "Dense tamarack-sphagnum bog, 1 mile south of Kent." These were evidently the first herbarium specimens collected from Kent Bog, the first scientific documentation of the plant life growing there.

Several years later, on April 21, 1968, I visited Kent Bog with Kent State graduate student Robert Hoffman. Bob had discovered the bog while hiking near his apartment on Sunnybrook Road and wanted me to see it. We entered from the west and, here also, went in only a short distance. I made a collection of gray birch (*Betula populifolia*) and on the label wrote "Sphagnum bog, two miles south of Kent." It was not until sometime later that I realized this site was a different part of the same bog I had visited in 1961.

On June 20, 1978, Kent State graduate student Barbara Andreas made collections of tamarack (*Larix laricina*) and leather-leaf (*Chamaedaphne calyculata*) at the site. Her later collections there included several species of sphagnum moss (*Sphagnum* spp.).

These plants are all still present in good numbers at Kent Bog today. The specimens are housed in the Kent State University Herbarium.

STATE NATURE PRESERVE

In 1983 a new line appeared on the Ohio Individual Income Tax Return form. On it those who had overpaid their tax and were entitled to a refund could indicate the part of their refund they wished to donate "to nature preserves, scenic rivers, and endangered species habitat protection." The response was generous.

On February 12, 1985, a large part of Kent Bog was purchased by the ODNR using these citizen-donated funds. The land was acquired from Harry and Olive Stark, who had owned and protected the bog since 1944. This was the first purchase under the new state program. The acquisition was officially designated Kent Bog State Nature Preserve and placed under the management of the ODNR's Division of Natural Areas and Preserves (DNAP). On February 27, 1987, the articles of dedication were filed with the Portage County recorder in Ravenna, giving the habitat and its rare plants a new degree of protection. The force behind the establishment of Kent Bog State Nature Preserve was DNAP assistant chief (later chief) Guy L. Denny, one of Ohio's great conservationists.

In May 2005, Gordon F. Vars, emeritus professor of education at Kent State, organized a group called Friends of the Kent Bog. Within a few years, under his leadership, the group grew to more than 500 members. They provide a valuable constituency supporting the Kent Bog State Nature Preserve.

THE BOARDWALK

In 1993 DNAP staff members and volunteers constructed a boardwalk, a 2,600-foot loop trail, that enables visitors to see all parts of the preserve without getting their feet wet or suddenly sinking up to their knees in wet peat. The boardwalk is constructed from planks of plastic lumber, the plastic having been recycled from such items as milk and soft drink containers. Emliss Ricks Jr., former preserve manager for DNAP, described

Gray birch trees, near boardwalk, Kent Bog State Nature Preserve (Photograph by Gerry Simon)

the walk in these words: "Dyed gray to resemble weathered wood, the recycled plastic boardwalk is not only visually and structurally acceptable, it is also aesthetically pleasing."[1] Names of contributors who supported the boardwalk, or of persons the contributors wished to honor, are carved in planks spaced at regular intervals.

The boardwalk is wheelchair accessible and has several turnouts allowing chairs to pass. Benches along the way provide places for rest and relaxing views of the bog.

The boardwalk starts at the parking lot. Going clockwise on the loop trail, it passes an information kiosk before crossing the wooded bank that surrounds the bog, goes down a small slope, crosses the moat, and enters the bog proper. For a short distance, the walk runs parallel to the moat, passing leather-leaf shrubs, blueberry shrubs, gray birch trees, and an occasional tamarack. A right turn takes the boardwalk into the large central core of tamarack trees. At places, a 360-degree sweep of the sky shows nothing but tamaracks.

Emerging from the trees, the walk goes through an extensive stand of blueberry shrubs. On the right it passes a clearing in the blueberry thicket, cut to promote growth of small cranberry plants. Farther along, the walk passes through a colony of leather-leaf, then crosses the moat and ascends a short rise to the bank. It proceeds along the bank through the hardwood trees and returns to the parking lot.

MASTODONS AND TURTLES

On the first plaque, at the start of the boardwalk, an artist's sketch shows two mastodons roaming in front of a glacier. Distant relatives of modern elephants, mastodons are now extinct worldwide. But in the past, such an event may actually have occurred here, the animals perhaps having come to the lake for water. Mastodon skeletons have not been found at Kent Bog, but they have been discovered at several Ohio sites similar to this.

Plaque number 11 describes some of the bog's current wildlife members that visitors may occasionally see. "Kent Bog is home to a small population of spotted turtles (*Clemmys guttata*). Named for their bright yellow spots, these secretive, palm-sized turtles are scarce in Ohio. To ensure that this boardwalk does not impair free movement of the turtles, 'turtle tunnels' fashioned from six-inch plastic pipe are placed at various intervals along the boardwalk. Other animals that may be encountered along the trail include green frogs, garter snakes, and the elusive smooth green snake. Kent Bog is also home to deer, foxes, raccoons, skunks, opossums, and cottontail rabbits. During the summer a number of birds including veeries, cedar waxwings, yellowthroats, and rufous-sided towhees are commonly heard and seen throughout the bog."

Although I have been to the bog many times, I have seen the shy spotted turtle on only a few occasions.

VISITORS

This area was purchased and preserved because of its scientific value and its role in Ohio's natural history. Hundreds of miles to the north, in Canada, this type of bog community is common, but in Ohio it is rare. Many of its plant species are, in Ohio, at or near the southern boundary of their range. Today, students and researchers come to study the bog and to learn the characteristics of its plant and animal life firsthand.

Other visitors come to the bog for the contact with nature it provides, in contained and relaxing conditions. Many also find pleasure in observing the patterns of the plant life. The bog plants tend to be freely branched and the bog community is inclined to be dense, which produces an endless variety of structural patterns that satisfy the mind.

Because of the bog's suburban location, the public has greater access to it than is the case at many other state-owned nature preserves. It is open to visitors daily. The explanatory plaques describe the highlights of the bog's natural history and make possible an informative, self-guided tour. The preserve is an asset to the community, both local and extended, and bears acquaintance well. Many area residents visit it repeatedly, each time seeing something new or gaining some new insight into nature or themselves.

CHAPTER 8

A WALK AROUND THE BOG

In his essay titled "Walking," the nineteenth-century American naturalist Henry David Thoreau wrote, "The West of which I speak is but another name for the Wild; and what I have been preparing to say is, that in Wildness is the preservation of the World."[1] The last eight words have become one of the most frequently quoted passages from Thoreau's writings.

Until a few years ago, I visited Kent Bog nearly every day, always more than 300 days of the year. I continue to visit it frequently. I often think of the Thoreau line when I am walking at the bog and observe the wildness of the shrubs in the dense undergrowth, the wildness of the white-trunked birch trees scattered here and there throughout the bog, and the wildness of the tamarack trees congregated in irregular groups at the bog's center, their horizontal limbs branched and branched again.

As a beginning graduate student, I rented a room in a private home in Iowa City. The other rooms were rented by graduate students majoring in music, all of whom were enrolled in a class in musical composition. In the course of the semester, they analyzed several small pieces and one major work, *Symphony No. 2* of the Finnish composer Jean Sibelius. They had a portable record player and a vinyl recording of the symphony that they played over and over. At any time of night or day, when I came into the house, I was apt to hear it being played.

Now, when I am at the bog in the autumn, I sometimes hear *Symphony No. 2* playing in my mind. The bog is surrounded by a bank where oaks and maples grow. As autumn arrives, the oak and maple leaves gradually begin to turn color and after a few weeks reach their bright final peaks of yellow, orange, and red. Similarly, in the Sibelius symphony, after a long gradual increase in tension and tempo, the music reaches a grand finale that opens with a majestic theme. The opening theme of the finale is followed by a fanfare, as if heralding the peak of fall color in the oaks and maples and other broad-leaved trees. The music slows and softens, and the colors in the leaves begin to fade. Then slowly the tension in the musical score returns, once again leading to the finale's majestic opening theme and, again, followed by the fanfare. This time the fanfare announces the peak of gold color in the thousands of tamarack needles. The music ends, and the leaves and needles fall.

The Seasons at the Kent Bog State Nature Preserve

A visit to Kent Bog is like a botanical trip to the north country. Tamarack and other boreal plants of the bog, such as leather-leaf, small cranberry, and tawny cotton-grass, grow as far north as Labrador and Newfoundland and, from there, range westward across Canada to the Yukon and on into Alaska. Their presence in Kent is a botanical wonder. The bog is accessible to visitors year-round.

The most remarkable plants are the tamarack trees, needle-bearing members of the pine family. Unlike pines and most other family members, such as firs, spruces, and hemlocks, the tamaracks shed their needles each year. In winter the trees are bare, their trunks and branches dark. In early spring, buds on the branches open and clusters of new needles emerge. The needles, soft to the touch and at maturity about a half-inch in length, remain green throughout the summer. In late autumn the tamarack needles turn yellow and fall, leaving the trees bare again.

Soon after the new needles emerge in spring, new cones also appear on the branches. On a large majority of trees, all the young cones are red,

Left: Young tamarack seed cones, red color phase. The circular tan structures are pollen cones. Kent Bog State Nature Preserve. (Photograph by Gerry Simon). Right: Young tamarack seed cones, rare green color phase. Kent Bog State Nature Preserve (Photograph by Gerry Simon)

but on a few all are bright green. The published descriptions of tamarack trees I have read mention only young red cones. The occurrence in this bog of young green cones is presumably genetic in origin, and perhaps is a local circumstance. Both red and green cones turn brown in midsummer. About a half-inch long, the brown cones remain attached to the tree for several years.

WINTER

In winter, Kent Bog is dark and usually quiet. Most of the woody plants are bare. The sedges and ferns and other soft plants are fallen and brown. Only the white trunks of the birch trees break the darkness.

Yet, along the preserve trail, a few patches of color are visible. The upper branches and buds of the blueberry shrubs are dark red. There is a greenish cast to the persisting brown leaves on the leather-leaf shrubs. The trunks of the tamaracks and other trees often have a crust of gray-green lichens at their base. When not covered by snow or dead leaves, the sphagnum moss is bright green across the floor of the bog.

A few shrubs of the deciduous winterberry holly grow here and there around the bog. Usually these shrubs produce only a few berries, but in occasional years their leafless branches are covered in early winter with hundreds of bright red holly berries. The berries fall in January, leaving the branches empty until spring.

SPRING

Sometime in April or May, the leather-leaf shrubs announce the coming of spring to Kent Bog. Remarkably, the shrubs' leaves, which were brown during the winter, slowly regain their green color. Then, overnight, thousands of small, white, urn-shaped flowers appear on the branches from flower buds formed the previous summer. Each flower is about one-quarter inch in length. The exact date of the event varies considerably from year to year, depending on the weather.

The bog has only a few conspicuous flowers. Nearly all are white, and nearly all open early in the growing season. At about the same time that the leather-leaf shrubs are blooming, the small, scattered serviceberry trees produce their flowers, each with five, narrow, white petals. Then the blueberry flowers open, whitish and urn-shaped like those of the leather-leaf, fellow members of the heath family. Last of the white flowers are those of the chokeberry shrubs, each with five nearly circular white petals. At the end of spring or in early summer, the small cranberry blooms. Its four reflexed roseate petals give each flower the appearance of a miniature red rocket.

Of the bog's flowers, only those of the blueberry shrubs have a fragrance, and that is only slight. Still, when tens of thousands of blueberry flowers are open at the same time, there is a noticeable sweetness in the bog air.

The bog is home to many species of small birds. The common black-capped chickadee is seen frequently flying among the trees. Occasionally

Leather-leaf flowers, Kent Bog State Nature Preserve (Photograph by Gerry Simon)

an eastern towhee, with its rufous-colored sides and white tail corners, flashes by. Other species of birds are generally seen less often.

For many who hear it, the song of the veery, a thrush, is the most prized sound in the bog. A secretive and usually hidden brown bird, it has a song Peterson describes as "liquid, breezy, ethereal."[2] McCormac and Kennedy write that the bird "sings a rapidly cascading series of sweet, ethereal notes that have an odd, haunting resonance as if sung within a pipe."[3] The veery's nearest relative is the hermit thrush, the "shy and hidden bird" of Walt Whitman's elegy for Abraham Lincoln: "When Lilacs Last in the Dooryard Bloom'd." The veery's carol floating over the bog on a morning in late April marks the bird's return to its northern home. Its nest, built on the ground or in a low shrub, has a few greenish-blue eggs.

As spring progresses, the young tamarack needles lengthen, and the broad leaves of the birch trees and the bog's other woody plants expand. The buried rhizomes (stems) of the ferns and sedges send up new green shoots and leaves. And then across the bog the silent wheels of photosynthesis begin to turn.

Trail through hardwoods on bank around Kent Bog (Photograph by Dick Bentley)

SUMMER

On a summer's day at Kent Bog State Nature Preserve, one sees mostly various shades of green, each shade coming from a different plant species. While in winter nearly everything in the bog is dark, in summer nearly everything is green. Other transitory colors appear as the season progresses.

In midsummer the wild blueberries ripen, the sweet, bluish-purple globes clustered on the ends of branches. Toward the end of summer and into autumn, red-capped mushrooms show up along the trail through the tamaracks, the broad cap borne on a short, white stalk.

Butterflies frequent open places, lighting briefly on one plant, then flying off to another. Two are striking in appearance. The wings of the red admiral are black and bordered with an orange-red stripe. The wings of the mourning cloak are dark brown and bordered with a row of blue spots above a yellow stripe.

Small garter snakes sometimes lounge on the boardwalk. Nonpoisonous and with dull yellowish stripes running the length of their dark bodies, they hurry away when someone draws near. Also nonpoisonous, the smooth green snake is not often seen. Camouflaged by its pale green color and elusive by nature, this gentle snake moves quickly out of sight at the least disturbance.

The bog's most celebrated wildlife resident is the spotted turtle. Like the veery and the smooth green snake, it is rarely seen. Small, as local turtles go, it is about four to five inches long and has yellow spots on its shell and body. The spotted turtle also is shy, quietly slipping into the dark water and disappearing when a hiker approaches.

In most summers, the mosquito population at the bog is small, the standing water being too acidic for their larvae. But in a wet summer, when the acid is diluted by the rain, the mosquitoes prosper.

On a hot day in August, the bog is humid, green, and still. Summer seems slated to go on forever.

Above: Tamarack trees in early November at Kent Bog State Nature Preserve. Blueberry shrubs (red leaves) in foreground. (Photograph by Gerry Simon)
Below: Hardwoods in fall color on bank around Kent Bog. Young green tamarack, left foreground (Photograph by Sue Cooperrider)

FALL

Shortly after mid-October, the Kent Bog State Nature Preserve reaches its first peak of fall color. On the bank encircling the bog, the green summer leaves of the maples, oaks, and other hardwood trees turn to shades of red, yellow, and orange. Down in the bog, the blueberry leaves are red, as are those of the stunted maples. The leaves of the gray birch trees are bright yellow.

Toward the end of October, or in early November, the preserve's second peak of fall color arrives, when the tamarack forest at the center of the bog becomes a sea of gold. After a few days of bright gold color, the tamarack needles turn brown and fall. Those landing on the boardwalk collect in temporary windrows on the gray planks. The needles falling into the bog settle on the green sphagnum moss.

The annual cycle of the seasons is completed. But the green moss holds the promise that the sequence of life will begin again.

AFTERWORD

BY DAVID E. BOUFFORD

The first time I learned of Tom Cooperrider was when I was a graduate student in the mid-1970s. My research was a worldwide study of a group of herbaceous plants, the Enchanter's-nightshades (*Circaea*), and Tom had written an account of the genus in Ohio, which he published in the journal *Rhodora* in 1962. Although only five pages long, because of the soundness of the research and the way it was presented so clearly and concisely, the contents of that paper have remained with me ever since.

In *Botanical Essays from Kent,* Tom's clear and concise writing style is again apparent. The essays cover nearly the entire last half of the twentieth century from Tom's own experiences. Also woven in are details from historical accounts dating back to the 1700s, which provide a sound background for his description of the natural beauty in the vicinity of Kent, Ohio. In covering the more than 200 years of study of the flora of Ohio, Tom provides the reader with mental images of Ohio as it was transformed over time to meet the needs of an expanding population and how it appears today. The essays' main focus, however, is the plants of Ohio—from the special ones, such as the state wildflower, *Trillium grandiflorum,* to weedy invasives such as garlic mustard, which lead local conservation groups to organize "garlic mustard pulls," and white mulberry to Labrador-tea and tamarack bogs—and their study and preservation.

Within the essays are accounts of mastodons and Pleistocene glaciers, spotted turtles, towhees, waxwings, deer, foxes, raccoons, and green snakes. There are readings from Thoreau and reminiscences of *Symphony No. 2,* by Jean Sibelius, plus a history of conservation and the preservation of natural areas in Ohio.

Despite the economy of space, these essays cover a wealth of information both from personal experiences and observations and from historic accounts. A single reading of these pieces will provide knowledge of the Kent area in general, but second and third readings will reveal even more that might have been overlooked the first time, such as the role and value of an herbarium in the environmental movement.

The *Botanical Essays from Kent* will appeal to a wide audience—not only to those interested in plants, Kent State University, or the state of Ohio, but to anyone interested in the history of the late twentieth century when many changes were taking place, particularly in the study of plants and in America's changing attitudes about conservation and the environment. It is clear to see Tom's love of plants and nature in his writings and to sense his appreciation of the people and places that have influenced and enhanced his life and continue to do so, all of which makes reading these essays so enjoyable.

DAVID E. BOUFFORD
Senior Scientist, Harvard University Herbaria

NOTES

Introduction

1. Rachel Carson, *Silent Spring* (Boston: Houghton Mifflin, 1962), 297.

2. Jim Clash, "Right Place, Right Time," *Explorers Journal* 87, no. 2 (2009): 30–33, quotation from 33.

3. *Record-Courier,* May 28, 2010, A4.

1. The Local Flora

1. Tom S. Cooperrider, Allison W. Cusick, and John T. Kartesz, eds., *Seventh Catalog of the Vascular Plants of Ohio* (Columbus: Ohio State University Press, 2001).

2. Morris Schaff, *Etna and Kirkersville* (Boston: Houghton Mifflin, 1905), 83.

3. Ibid., 89.

4. Michael B. Lafferty, ed., *Ohio's Natural Heritage* (Columbus: Ohio Academy of Science, 1979).

5. E. Lucy Braun, *Deciduous Forests of Eastern North America* (Philadelphia: Blakiston, 1950).

6. George W. Knepper, *An Ohio Portrait* (Columbus: Ohio Historical Society, 1976), 61.

7. *Akron Beacon Journal*, May 16, 2010, B1.

8. Jim McCormac and Gary Meszaros, *Wild Ohio* (Kent, Ohio: Kent State University Press, 2009).

9. Cooperrider, Cusick, and Kartesz, *Seventh Catalog.*

10. Karl H. Grismer, *The History of Kent* (Kent, Ohio: Record, 1932).

11. Donald Wyman, *Trees for American Gardens* (New York: Macmillan, 1965).

12. Tom S. Cooperrider, *The Dicotyledoneae of Ohio,* part 2, *Linaceae through Campanulaceae* (Columbus: Ohio State University Press, 1995).

2. The Floristic Botanists

1. Ronald L. Stuckey, "Early Ohio Botanical Collections and the Development of the State Herbarium," *Ohio Journal of Science* 84 (1984): 148–74.

2. George W. Knepper, *An Ohio Portrait* (Columbus: Ohio Historical Society, 1976).

3. Jay F. Laning, "The Evolution of Ohio Counties," *Ohio Archaeological and Historical Publications* 5 (1897): 326–50.

4. Harry F. Lupold and Gladys Haddad, eds., *Ohio's Western Reserve: A Regional Reader* (Kent, Ohio: Kent State University Press, 1988).

5. Tom S. Cooperrider, "Ohio's Herbaria and the Ohio Flora Project," *Ohio Journal of Science* 84 (1984): 189–96.

6. Tom S. Cooperrider, "Ohio Floristics at the County Level," *Ohio Journal of Science* 61 (1961): 318–20.

7. John S. Newberry, "Catalogue of the Flowering Plants and Ferns of Ohio," *Annual Report Ohio State Board of Agriculture* 14 (1860): 235–73.

8. Tom S. Cooperrider, "Changes in Knowledge of the Vascular Plant Flora of Ohio, 1860–1991," *Ohio Journal of Science* 92 (1992): 73–76.

9. Cooperrider, Cusick, and Kartesz, *Seventh Catalog.*

10. Charles C. King, ed., *A Legacy of Stewardship: The Ohio Department Natural Resources, 1949–1989* (Columbus: Ohio Department of Natural Resources, 1990).

11. Tom S. Cooperrider, ed., *Endangered and Threatened Plants of Ohio.* Ohio Biological Survey Biological Notes, no. 16 (Columbus: Ohio State University, 1982).

12. Ohio Division of Natural Areas and Preserves, *Rare Native Ohio Plants: 2008–2009 Status List* (Columbus: Ohio Department of Natural Resources, 2008).

13. Ohio Division of Natural Areas and Preserves, *Directory of Ohio's State Nature Preserves* (Columbus: Ohio Department of Natural Resources, 1996).

3. The Herbarium

1. E. Lucy Braun, *The Woody Plants of Ohio* (Columbus: Ohio State University Press, 1961).

2. Merritt L. Fernald, *Gray's Manual of Botany,* 8th ed. (New York: American Book, 1950).

3. G. Ledyard Stebbins, *Variation and Evolution in Plants* (New York: Columbia University Press, 1950).

4. Robert F. Thorne, "The Vascular Plants of Southwestern Georgia," *American Midland Naturalist* 52 (1954): 257–327.

5. Lawrence J. Eilers and Dean M. Roosa, *The Vascular Plants of Iowa* (Iowa City: University of Iowa Press, 1994).

6. Tom S. Cooperrider, *The Ferns and Other Pteridophytes of Iowa.* Iowa Studies in Natural History (Iowa City: State University of Iowa, 1959).

7. Tom S. Cooperrider, "On the Value and Feasibility of a Plant Distribution Atlas for the States in the Gray's Manual Range," *Ohio Journal of Science* 88 (1988): 84–86.

8. Edgar Anderson, *Introgressive Hybridization* (New York: Wiley and Sons, 1949).

9. Miwako Cooperrider, "Introgressive Hybridization between *Quercus marilandica* and *Q. velutina* in Iowa," *American Journal of Botany* 44 (1957): 804–10.

10. G. Ledyard Stebbins, "Concepts of Species and Genera," in *Flora of North America North of Mexico,* vol. 1, *Introduction* (New York: Oxford University Press, 1993).

11. Braun, *Woody Plants of Ohio.*

12. J. Arthur Herrick, *The Natural Areas Project: A Summary of Data to Date.* Informative Circular no. 1 (Columbus: Ohio Biological Survey, 1974).

13. Robert B. Gordon, *Natural Vegetation Map of Ohio at the Time of the Earliest Land Surveys* (Columbus: Ohio Biological Survey, 1966).

14. Robert B. Gordon, *The Natural Vegetation of Ohio in Pioneer Days* (Columbus: Ohio Biological Survey, 1969).

15. Cooperrider, "Ohio Floristics at the County Level," 318–20.

16. J. Lanjouw and F. A. Stafleu, *Index Herbariorum,* part 1, *The Herbaria of the World,* 4th ed. (Utrecht: International Association for Plant Taxonomy, 1959).

17. Tom S. Cooperrider, *The Vascular Plants of Clinton, Jackson and Jones Counties, Iowa.* Iowa Studies in Natural History (Iowa City: State University of Iowa, 1962).

18. Robert F. Thorne and Tom S. Cooperrider, "The Flora of Giles County, Virginia," *Castanea* 25 (1960): 1–53.

19. Miwako K. Cooperrider and Tom S. Cooperrider, "History and Computerization of the Kent State University Herbarium," *Ohio Journal of Science* 94 (1994): 24–28.

20. Tom S. Cooperrider and Clinton H. Hobbs, "Almon Rood: The Western Reserve's Botanist," *Western Reserve Magazine* (July–August 1978): 18–20, 50–52.

21. Heidi Hetzel-Evans, "A Lifetime Devoted to Ohio's Flora," *Natural Ohio* 27, no. 3 (2005): 5–7.

22. American Society of Plant Taxonomists Advisory Committee, *Systematic Botany Resources in America* (New York: New York Botanical Garden, 1974).

23. Ibid., 1.

24. Kent Historical Society Book Committee, *Kent, Ohio: The Dynamic Decades* (Kent, Ohio: Kent Historical Society, 1999).

25. In the following decade, two endowments were established to provide future support for the herbarium, the Tom S. and Miwako K. Cooperrider Herbarium Endowment Fund, founded in March 1992, and the J. Arthur Herrick Herbarium Endowment Fund, founded in October 1997.

4. A Colleague

1. Lupold and Haddad, *Ohio's Western Reserve*, 33.

2. Herrick, *Natural Areas Project.*

3. J. Emerson Kempf and J. Arthur Herrick, "Effect of Penicillin on Experimental Staphylococcus Osteomyelitis in Rats," *Proceedings of the Society for Experimental Biology and Medicine* 58 (1945): 100–102.

4. Herrick, *Natural Areas Project*, 1.

5. Ibid., 1.

6. Ibid., iv.

7. Ibid.

5. The Campus Plants and Gardens

1. Ruth Rogers Clausen and Nicholas H. Ekstrom, *Perennials for American Gardens* (New York: Random House, 1989).

2. In a 2010 phone conversation, Judith Beyer Murin's sister, Janet, recalled seeing a plaque with Judith's name beside a small garden during a campus visit in 1975.

3. *Record-Courier,* April 23, 2009, A1.

4. Flora of North America Editorial Committee, eds., *Flora of North America North of Mexico,* vol. 3, *Magnoliophyta: Magnoliidae and Hamamelidae* (New York: Oxford University Press, 1997).

5. Bailey Hortorium Staff, *Hortus Third* (New York: Macmillan, 1976).

6. Cooperrider, Cusick, and Kartesz, *Seventh Catalog.*

6. A Specimen Tree

1. Bailey Hortorium Staff, *Hortus Third,* 344.

2. Ibid., 529.

3. Fred C. Galle, *Hollies: The Genus Ilex* (Portland, Ore.: Timber Press, 1997), 353.

4. Cooperrider, *Dicotyledoneae,* 66.

5. Galle, *Hollies,* 353.

6. Cooperrider, *Dicotyledoneae,* 66.

7. The Bog

1. Emliss Ricks, "Look What's Happening at Kent Bog State Nature Preserve," *ODNR/DNAP Newsletter* 15 (1993): 2.

8. A Walk around the Bog

1. Henry David Thoreau, *The Works of Thoreau,* ed. Henry Seidel Canby (Boston: Houghton Mifflin, 1946), 672.

2. Roger Tory Peterson, *A Field Guide to the Birds of Eastern and Central North America* (Boston: Houghton Mifflin, 1980), 222.

3. James S. McCormac and Gregory Kennedy, *Birds of Ohio* (Edmonton, Canada: Lone Pine Press, 2004), 234.

INDEX TO PLANT NAMES

Page numbers of photographs are in **boldface**.

INDEX

Page numbers of photographs are in **boldface.**

DATE DUE

GAYLORD

PRINTED IN U.S.A.

QK 180 .C66 2010

Cooperrider, Tom S.

Botanical essays from Kent